MATH
ANXIETY

STRATEGIES TO INCREASE CONFIDENCE
IN YOUR STUDENTS WHO FEAR MATH

Other books by Brett Bernard

Total Math Engagement

How to Talk Math

MATH
ANXIETY

STRATEGIES TO INCREASE CONFIDENCE
IN YOUR STUDENTS WHO FEAR MATH

by

—— America's Chief Math Officer™ ——

Published by Gold Medal Staff Development, LLC

www.brettbernard.com

Copyright © 2016 Gold Medal Staff Development, LLC

All rights reserved

In accordance with the U.S. Copyright Act of 1975, the reproducing, scanning, uploading, and electronic sharing of any part of this book without permission of the publisher is unlawful piracy and the theft of the author's intellectual property. If you would like to use material from the book (other than for review purposes), prior written permission must be obtained from the publisher. Thank you for supporting the author's rights.

Cover Design: Pro Designs

Printed by CreateSpace

Printed in the United States of America

ISBN-13: 978-1537222509

ISBN-10: 1537222503

Dedication

This book is dedicated to my kids, Brody and Nora. My hope is that teachers and parents can use the strategies and examples in this book, so my kids, and all other kids, never need to have a negative experience in math. My hope is that all students will use these strategies to cope with any challenging math situations. Daddy (August, 2016)

Table of Contents

Chapter 1: What Is Math Anxiety?......................1

Chapter 2: Causes of Math Anxiety 11

Chapter 3: Strategies for Decreasing Math Anxiety 21

Chapter 4: Additional Resources..................... 43

About the Author... 61

Chapter 1:
What Is Math Anxiety?

"Believe in yourself. Have faith in your abilities! Without a humble but reasonable confidence in your own powers you cannot be successful or happy."

Norman Vincent Peale

I hated math most of my life. I did not appreciate it or care about it until I overcame my fear and anxiety of math and going to math class. I AM NOW A MATH ANXIETY SURVIVOR! I was inspired to write this book after sharing my story with a group of students and teachers who had the same problem I had. By sharing my experiences and

the strategies I used to overcome this fear and anxiety, they all became math anxiety survivors.

There are some things in life that we never forget. They can be either positive or negative experiences. Either way, we often let those experiences define us. When I was a kid, I had a lot of positive experiences in school and in athletics. One example that comes to mind is that one of my teachers told me that I had a gift for public speaking. He said this after I gave an impromptu speech in front of my class. I never forgot that. It made me want to please him even more. To this day, I am grateful to Mr. Studer for making speech class a positive experience for me and inspiring me to always improve. I never forgot that moment.

I had always struggled with math. I worked hard on it, but often I didn't get it. I wanted to do well. But, to be honest, it was just hard for me. In the fifth grade, my teacher would start the lesson each day by writing some math exercises on the board. She would then choose five students to go up in front of the class and do them.

Chapter 1: What Is Math Anxiety?

I was always nervous to be called on. I would try to get out of the classroom to use the bathroom, or get a drink of water, or any other way I could think of. It never worked. I got chosen to go up in front of the class all the time. I would shake, I would turn beet red, my entire body would get warm, and I would feel sick. The other kids would laugh at me. It made me feel stupid. I let the moment define me. I began to hate math, and I never forgot that moment.

There were many other moments when I felt like that, but the math class sealed the deal. I hated math. Instead of asking to use the bathroom to get out of class, I began to get in trouble, because I knew I would have to leave the room. I hated math most of my life. As I got older, I always thought back to that moment at the chalkboard. I still think of being that kid in front of the class when I am teaching my students. I can tell you from experience that math anxiety is real. I have experienced it, and I have survived it.

Maybe you've had an experience in your life that has forced you to change how you live? perhaps a fear of animals? a fear of heights or small places? maybe a fear of being in an airplane? Regardless of the fear, you know how you feel when you're put in a situation where you're scared or not comfortable. Maybe you shake. Maybe you become warm and turn red, or you lose your breath! Now just imagine feeling this way every single time you go to a math class. How can a child possibly learn if they have these symptoms?

> **"I can tell you from personal experience that math anxiety is real. I have experienced it and survived it."**

Mark Ashcroft defines math anxiety as a feeling of apprehension or fear that interferes with math performance. In other words, if someone has math anxiety, they have a hard time learning. There are many consequences of math anxiety including a fear of taking advanced math classes, being unwilling to attempt challenging math exercises, and being unusually nervous in math class.

According to Perina (2002), "Math anxiety hinders students' working memory. It occurs at different ages in different people for different reasons." Some students have reported having panic attacks in math class. According to Buxton (1981) "Panic can be seen as a turbulence in the mind, a kind of mental frenzy. The mind may freeze and the student may experience physical tension and rigidity." I remember feeling like that in my math classes. I froze. I couldn't think. Time stood still. Again, think of a time when you have been anxious or scared and imagine what it is like for the kids in your classroom who fear math.

Symptoms of math anxiety can be physical and psychological and may include the following:

Physical: Nausea, shortness of breath, sweating, heart palpitations, increased blood pressure.

Psychological: Memory loss, paralysis of thought, loss of self-confidence, negative self-talk, math avoidance, isolation (thinking they are the only one who feels this way).

Source: Preis, Christy, Biggs, Bobbi T. (2001) *Can Instructors Help Learners Overcome Math*

Anxiety? ATEA Journal, Volume 28(4), 6-10, Apr/May.

Things to know about math anxiety

1. "The long-term effects of math anxiety can lead to avoiding courses, college majors, and career paths that involve mathematics." (Ashcroft and Moore, 2009)

2. "Students who suffer from math anxiety feel alone, helpless, and often these feelings exacerbate anxiousness a student feels while learning mathematics." (Tobias, 1993)

3. "Children do not appear to be troubled by math anxiety while in the first two or three years of elementary school." (Ashcroft and Moore, 2009)

4. "A student's ability to concentrate while learning and doing math work is essential to being successful in the area of mathematics." (Tobias, 1993)

Chapter 1: What Is Math Anxiety?

As I suffered through very high levels of math anxiety, I found many ways to avoid math. I had always been a very well behaved student in class, but that changed quickly. My misbehavior resulted in me sitting in the hallway or in the corner of the room for a time out. As I look back on my education, I can make the connection that my anxiety also resulted in missing out on learning. Getting out of doing math and avoiding math left me unprepared. This increased my anxiety even more. It became a cycle. I avoided math and became unprepared. When it came time for assessments, I felt like even more of a failure.

Some symptoms of math anxiety

- Nervous and unable to focus in math class
- Fear of starting a math exercise or problem that appears difficult
- Dread going to math class
- Feeling embarrassed, irritated, or frustrated
- Fear of answering a question incorrectly
- Being nervous of peer reaction

Math Anxiety Cycle
anxiety → avoidance → unprepared → more anxiety

Math anxiety can happen to anyone, regardless of achievement level. I have worked with many students in my career who get math. They love it, and they are exceptional. However, there is a lot of pressure when you're working at an advanced level. There is an expectation that you should always do better, and you should always do more. Getting all "A's" in third grade math is not good enough anymore. There is now pressure to move up a grade level or get private tutoring to advance even further.

Regardless of age, grade level, or ability level, math anxiety is real. Each student and teacher should take the quiz on the next page.

Chapter 1: What Is Math Anxiety?

Do You Have Math Anxiety?

This is a test where you rate yourself. Rate your answers from 1-5, add them up, and check your score below.

1= Disagree   5=Agree

1.	I cringe when I have to go to math class	1 2 3 4 5
2.	I'm uneasy about going to the board in a math class	1 2 3 4 5
3.	I'm afraid to ask questions in math class	1 2 3 4 5
4.	I'm worried about being called on in math class	1 2 3 4 5
5.	I get math now but worry that it will be difficult soon	1 2 3 4 5
6.	I tend to zone out in math class	1 2 3 4 5
7.	I fear math tests more than any other kind	1 2 3 4 5
8.	I don't know how to study for math tests	1 2 3 4 5
9.	It's clear to me in math class, but I don't get it at home	1 2 3 4 5
10.	I'm afraid that I can't keep up with the class	1 2 3 4 5

Check your score

40-50 — High anxiety

30-39 — Anxious

20-29 — Low anxiety

10-19 — Wow! A math major in the making!

Source http://www.mathpower.com/anxtest.htm

Chapter 2:
Causes of Math Anxiety

"Put your heart, mind, and soul into even your smallest acts. This is the secret of success."
>   Swami Sivananda

According to Rising (October 2013), parents are the main cause of math anxiety. In my experience, teachers and society are the other reasons that kids experience this. This book will primarily focus on the parent and teacher aspect that contributes to math anxiety.

Math Anxiety

Three causes of math anxiety

1. Parents
2. Teachers
3. Society

How often have you met with parents at conferences and heard these statements?

- "We can help her until she gets to fifth grade, but after that, she's on her own."

- "I wasn't good at math, so I don't think my son will be either. That's okay."

- "Our daughter is just not a math person."

- "His sister is in accelerated math, so he should be too."

- "My fourth grader is doing the same math that I did in eighth grade. I can't help him."

- "He knows all of his math facts. Why can't you put him in accelerated math?"

As teachers, we have all heard the saying "the apple doesn't fall far from the tree." These statements are just one example of how well-intentioned parents can contribute to math anxiety. Many parents think that if they weren't good at math, their child won't be either, and the student usually receives this message. I have worked with hundreds of families over my career, and I have never seen a situation where knowledge of math is hereditary.

In today's world, kids learn differently than most of their parents did. It's a new age. Teachers use innovative methods that parents are often not familiar with.

When I was a student, I was expected to memorize my math facts and just know them. Now, teachers use multiple strategies that parents can't always relate to when helping their child—things like fact families, partial product multiplication, using the lattice method to multiply, and many others. I have seen multiple examples of families struggling to help their kids because the parents don't understand how to help. There is a new language in math and parents don't always know how to talk math with their kids. Having a common vocabulary in this area is critical for success.

There is a rule of thumb in the math community that a child should not be placed in accelerated math until they are in the seventh grade. The rationale for this is that by skipping a grade, like third or fourth, the students are missing out on foundational skills and procedural fluency. I have taught accelerated math and seen the tears coming from my students who have amazing math minds but don't know how to do a long division problem.

There are always exceptions, and there are students who consistently score in the top percentile, but overall, most students benefit from enrichment in math class versus acceleration. I know many parents who push their kids to be in accelerated math. Their kids are intelligent but not quite ready to be accelerated. The best thing for kids in that situation is to thrive in their math classroom and benefit from enrichment activities, problem-based math, and project-based math. I have seen some great math minds withdraw from math because of constant pressure and the message that what they are doing is never good enough.

All parents want their children to succeed, and it is imperative that the teacher give them the tools to do this. It takes patience, support, and a positive attitude.

Teachers play a big role in the level of anxiety their students have in math class. Think about it, if you have a bad attitude about something, others around you likely will. "Teachers and parents who are afraid of mathematics pass that on to their children." (Furner, 2003)

* * *

Educators of all backgrounds and levels of experience can unknowingly do many things that can cause math anxiety in their students. Teachers need to review foundational mathematical skills with their students. This is just like coaching an athletic team at any level. Professional basketball players start each practice doing drills, lay-ups, free throws, and passes. Math teachers need to follow this model as well because students need to master the basics before they move on to more complicated problems. This is especially important for advanced students, because it gets them warmed up for math class, and it is a very welcome brain break when working on complex problems. A little bit of maintenance goes a long way.

Cultural and gender equity is another reason students experience math anxiety. Many teachers send the message that girls cannot perform as well in math (Jackson and Leffingwell, 1999). Teachers who promote this idea are one of the reasons that many girls give up without trying.

Elementary teachers in the United States are almost exclusively female (90%). Evidence tells us that female teachers often pass on a gender stereotype that it is acceptable for a girl not to be good at math. In fact, a study that included seventeen mid-western teachers and their students tells us that the anxiety of the teachers relates to girls' math achievements via girls beliefs about who is good at math.

The teachers' math anxiety and the math achievements of the students in their classrooms were also assessed. The scores indicate that there was no relation between the anxiety of the teacher and the achievement of her students at the beginning of the school year. But, by the end of the school year, the higher the teacher's anxiety about math, the more likely the girls were to believe that they were good at reading and the boys were good at math.

Additionally, girls who believed this stereotype had significantly worse achievements than girls who did not. In elementary school, where most teachers are female, the math anxiety of a teacher has consequences for the achievement of the girls by influencing their beliefs about who is good at math. There was no significant relation between the teacher's anxiety and the achievements of the boys in the classes. [Sean Beilock, Elizabeth Gunderson, Geraldo Ramirez, Susan Levine (2009).]

Some examples of how teachers may cause math anxiety

- Not caring about their students

- Not caring about math

- Being angry or frustrated when students don't understand a concept

- Setting unreasonable expectations

- Using math as a punishment

- Not using a variety of teaching methods

- Gender bias

- Cultural bias

- Focusing on the correct answer and not the process

- Using traditional or old-fashioned methods

Our society places a lot of pressure on students and teachers, and society often sends messages that affect how students perceive math. I have seen many movies where highly intelligent girls are put in situations where they pretend to be bad at math to get a boy to like them or to fit in with a peer group. This message supports the multiple studies of gender stereotypes in math. We are constantly hearing how students in the United States compare to other countries in math, and how we put more emphasis on testing rather than conceptual understanding.

> **Reasons society plays a role in causing math anxiety**
>
> - Gender and cultural stereotypes in the media
> - Politicians sending the message that what is being done is not good enough
> - Comparing the test scores of American students to other high-achieving countries
> - Negative talk about our economy
> - Excessive amounts of standardized testing
> - Some states publish test scores (by school) in newspapers

Chapter 3:
Strategies for Decreasing Math Anxiety

"Start by doing what is necessary, then do what is possible, and suddenly you are doing the impossible."

Francis of Assisi

We know the three main causes of math anxiety. Parents, teachers, and society have a huge influence on the lives of our students, and it is critical that steps be taken that will prevent and decrease math anxiety in all of our students.

> **Let's review the three causes of math anxiety**
>
> 1. Parents
> 2. Teachers
> 3. Society

When I was in elementary school, middle school, high school, and college, there were no strategies in place to help me decrease my high levels of math anxiety. The perception was that I should just study harder. It got to the point where just looking at a page in a math book was like trying to read it in a different language. I can think of many times when I studied really hard for a test. I did all of the work. I listened in class, and I thought I was ready for the test. I would then look at the test and wonder if I was in the right classroom. The test looked impossible, and I froze. Often, I just guessed, because I had no idea what to do.

* * *

We need to offer help to any student who is struggling in math and experiencing math anxiety. Following are some strategies that teachers can use to lessen their students math anxiety.

- **Talk about math in a positive way**

    "Teachers with math anxiety or a negative view of math contribute to the development of math anxiety in their students." (Sparks, 2011) Regardless of your experiences in math and how you view it, every teacher should talk about math in a positive way.

- **Promote a climate of equity**

    Many teachers send the message that girls do not perform as well as boys in math. This causes girls to give up without trying. All teachers need to give the same opportunities to boys and girls, and they should encourage all students.

- **Connect math to the lives of the students**

    Once we learn the experiences of each student and assemble their learning profile, we will know the whole child. This information should be used often to connect math to their lives. Simply using examples from their sporting events, music concerts, or other after school activities can go a long way. I once had a student share with me the total points he got in each of his basketball games. We used this information to find mean, median, mode, and range. Primary grade levels could have ordered the number or added them.

- **The process is more important than the product**

Math is usually perceived as a right or wrong subject. You are either correct or you are not. The emphasis needs to shift from only having one correct answer to looking at the computational process. I was correcting a third grade math test and the student got every question wrong. It was frustrating to me, so I looked at the test in more detail and saw that he really had an understanding of what he was doing. He was just making a simple mistake that resulted in him getting the question wrong. By looking at the process he used, I was able to give the support he needed and use that to drive my instruction.

Chapter 3: Strategies for Decreasing Math Anxiety

- **Have a positive discourse**

    Who is doing the talking in the math classroom? I learned more about math during my first year of teaching than in years of education. I was the only one doing the talking in my class. The ones doing the talking are the ones doing the learning. Often times, students become anxious, because they fear speaking in front of a large group or are concerned that they will get an answer wrong in front of their peers.

    Strategically, assembling small groups in your math classroom can go a long way in decreasing anxiety. This smaller setting is a more intimate way for students to participate, and the ones who are more reserved begin to emerge as stronger participants as the year progresses. "Cooperative groups provide students with opportunities to exchange ideas, ask questions freely, verbalize their thoughts, justify their answers, and debate processes." (Geist, 2010)

    Encourage the students to talk with each other. "What did you get for question number three, and how did you get it?" "I got a different answer." "Let's compare our work." There does need to be instruction from the teacher to begin the lesson along with monitoring and reviewing throughout the class. Once the students are ready ... let them start talking.

I keep my students in the same groups for at least ten days. This builds a level of trust for those who are more reserved and anxious, and they slowly begin to participate. Please refer to the Bernard Discourse Model™ in Chapter 4.

- **Promote the development of problem solving skills in girls**

    Girls generally solve problems using procedures they have been taught. Teachers should give girls the opportunity to find their own ways to solve a problem and praise them for doing so.

- **Use human resources**

    People are some of the best resources teachers can use in their classrooms. Seek out parents, community members, and other professionals of all cultural backgrounds to speak in your math classrooms. Make it a point to invite successful, confident women who use math in their careers. If you are unable to locate volunteers, look online for appropriate videos to share with your classes that portray women and math in a positive light.

- **Don't give homework ... give work time**

Yes, you read that correctly. Don't give kids traditional homework. I have a few reasons for this. First, math homework often causes a power struggle between students and their parents. There are times when ten minutes of homework turns into an hour-long struggle between the child and their parents. A common reason for this is that students are using different strategies from what their parents used. Other times, the well-intentioned parents just don't know how to help their child.

My other reason for not giving homework is that the students might not have mastered the concept. For example, a third grade teacher may have explained the steps to lattice multiplication (or any other method) and given their class twenty exercises to compete at home. If the student doesn't understand this, it leads to more frustration, incomplete work, or getting the incorrect answers. The result of this is negative—either the student receives a consequence or the teacher has to re-teach the concept.

Don't get me wrong, if a student who has the ability to complete their work and does not due to refusal or laziness, a plan needs to be put in place for him/her to complete the required work.

An alternate approach to assigning homework is to assign "work time." Kids should work hard during the day and have opportunities to get help from teachers and support staff to complete their work. There are multiple options for working on math at home. These include

Math work time options

1. Practice fact fluency
2. Watch online math videos
3. Play cards or board games as a family
4. Keep a math journal
5. Cook meals as a family
6. Look for math in the sports you play and watch; play fantasy sports as a family
7. Compete review worksheets of previously mastered material
8. Clip coupons from the newspaper
9. Use the book *How to Talk Math* available on Amazon
10. Measure the area/perimeter of each room in your home
11. Record the high and low temperatures each day
12. Exercise and use apps to record distance and time
13. Graph amount of time read each day
14. Plan math scavenger hunts
15. Get the FREE game "Choose Your Challenge" at www.brettbernard.com

- **Make them good at something**

    Self-efficacy is connected to a success or a failure with a math concept or task. If a student does well on a math task, their confidence increases and anxiety decreases. Conversely, when a student fails on a math task, their anxiety increases and confidence decreases.

    Teachers need to give their students opportunities to succeed when doing math. We need to set them up for success, so they can experience what it is like. If a student is not good at anything in math, make them good at something. Connect things like sports and other activities to math. Practicing and getting good at hockey takes the same amount of work as getting a good score on a math test. I used to complain about math and how I didn't get it. My mom would say, "If you can ride a unicycle, you can do math." She was right. We just need to encourage a transfer of skills such as the process of learning to play an instrument or a sport and apply them to math.

Chapter 3: Strategies for Decreasing Math Anxiety

- **Incremental release of information (IRI)**

    How many times have you become frustrated when you've purchased something that you have to assemble? Oftentimes, it is a massive page of information that is difficult to read. The print is small, all the information is on one page, and it is hard to follow. We can't bombard our math students with information for two reasons. First, it can cause anxiety (just like the directions to assemble something). Second, it can get the students off task because they're reading the information rather than listening to the teacher. Rather than putting every single step on the board beforehand, it is more effective to write the steps as you go. If I'm doing a PowerPoint presentation, I only have one step on the screen at a time. By the end of the lesson, my students can clearly see each step.

- **Have fun!!!**

    This is key. Laugh with your students. Smile. Talk about math in fun ways. Tell silly/appropriate math jokes. Have your students write math songs and share them with the class. Do math dances using geometrical vocabulary like parallel arms, right angles with your fingers, and making intersecting lines. Use cartoons to introduce a concept. Laughter relaxes us and makes us more comfortable.

More strategies teachers can use to reduce math anxiety

- Use the multiple intelligences (see information in Chapter 4)

- Assess students in multiple ways

- Use technology and other available resources

- Use manipulatives

- Prepare them for standardized tests

- Create an environment where milestones are celebrated

Chapter 3: Strategies for Decreasing Math Anxiety

What about the student who just can't overcome their math anxiety? There are those who need more of a long-term solution to become a math anxiety survivor. The key to this is to build trust. In my book, *Total Math Engagement*, I outline the E.P.I.C. process to achieve math engagement. I can tell you that it starts with knowing the whole child and building trust. It takes time, and it takes a lot of patience. It is an opportunity to establish a relationship with that child.

The goal is that they start talking about their math history with you. The question you need them to answer is "When was the last time you liked math, and why did you like it then?" This is the only information you will need. If the student you are working with is in the fourth grade and they tell you that the last time they liked math was in the second grade, you need to find out why. Maybe that teacher did something special, or they worked in small groups, did projects, or got to work independently. The important thing to do now is to go back to where they liked math.

Start teaching that child the way their second grade teacher did. If that teacher used manipulatives and you don't use them, now is the time to start. If a teacher can differentiate and accommodate for that child, that is the start to reducing and eliminating their anxiety and fear of math. I encourage teachers to take their time and be patient with this. The goal is to get these students comfortable in math class, and if you can accomplish that, then you are E.P.I.C.™!

* * *

Parents also have a huge impact on how their child perceives math. Following are some strategies that parents can utilize to decrease their child's math anxiety.

- **Talk positively about math**

    Math anxiety typically starts at around the third or fourth grade. However, the messages that students hear about math early on can be a huge factor in this. Think about it. When adults talk about numbers, what do kids hear? They hear us complaining about credit card bills, debt, and taxes. They begin to associate math with frustration. Parents and teachers need to talk about math in a positive way to send a message that math is something that we use every day, and that it can be fun. This could be done by playing games as a family, cooking, or showing how math can be used successfully.

- **Don't pressure them**

   My wife and I were working with our realtor to find a new house a few years ago. He learned that we were both teachers and commented that whenever he showed a family a new area that ninety percent of the parents asked what type of gifted education program the local schools had.

   As teachers, we know that every single child does not belong in a high potential class. All children have talents and are strong in certain areas, but the reality is that well-intentioned parents often set high expectations. According to Cavanaugh (2007), "Parents increase their child's math anxiety when they have unrealistically high expectations for their child's success. Please refer to the survey, *What Are You "Smart" At*, in Chapter 4 to assist you in identifying the strengths of each child.

- **Become involved in their math journey**

    Communicate with your child's teacher, read their classroom newsletters, look through your child's math book, and get to know their curriculum. If you notice math anxiety, contact your child's teacher to go over strategies to help your child. Work with your child when they need help but don't do the work for them. If you don't know how to do something, find someone who can help you. There is a list of resources in Chapter 4 that may be of assistance. Be patient and positive.

- **Practice fact fluency together**

    The expectation is that each student should know their basic math facts (addition, subtraction, multiplication, and division) within three to five seconds. If they are asked what 3x5 equals, they should say 15 within five seconds. Some students thrive on the competitiveness of timed tests, while many do not. Practicing fact fluency as a family can be helpful and fun. Some examples include flash cards, verbally asking, online games, and fun worksheets. Let your child quiz you. The important thing is to focus on improvement and not on speed. Speed and fluency will come with practice.

* * *

It is important to equip students with the tools they need to overcome math anxiety. There will be situations in their lives where they may have an anxiety attack because of a certain teacher, experience, math content, or influences from society. The following strategies will help students cope with math anxiety without help from teachers, parents, or others.

- **Visualize success**

    "If you want to reach a goal, you must see the reaching in your mind before you actually arrive at the goal."

    Zig Ziglar

This thinking can apply to all areas of life. I have visualized myself succeeding in academics and athletics. Every time I competed on my unicycle, I imagined myself succeeding, and every time I struggled with math, I imagined myself doing well. This is a powerful tool!

- **Students should believe in themselves**

All students should have a positive attitude. Remind them of things they are good at, like playing an instrument or being in a sport. Have them think about all the hard work they had to put in to succeed, and that they have to do the same with math.

- **Practice the basics**

Students should warm-up their math mind every single day. They should practice fact fluency. Their goal should be to do each basic math fact within five seconds. They can practice online, use flash cards, and play games.

Chapter 3: Strategies for Decreasing Math Anxiety

- **Learn math vocabulary**

    Every student needs to learn to talk math, and they should take the time to learn the key vocabulary words in math. This will help them to understand the teacher and significantly improve scores on tests. The book *How to Talk Math* is a very helpful resource and is available on Amazon.

- **Ask for help**

    There really is no such thing as a dumb question. Students should ask their teachers, parents, friends, and anyone else they trust if they have a question. Online videos are another excellent resource.

- **Call a friend**

    Whenever I'm feeling down, it has always helped me to call a friend. It sounds simple, but it is incredibly helpful. Let's say a teacher, supervisor, classmate, or coworker puts you down for whatever reason. You might be feeling anxious and frustrated. Pick up the phone and call a friend who you know sees the good in you and talk with them. I have done this many times, whether it involves math, a work experience, or something personal. The power of hearing someone say nice things about you is amazing.

Please share these strategies with your students so they can use them to cope when having trouble in math. I have found it helpful to list these in my classroom as a gentle reminder to my students.

* * *

Math anxiety is real, and it can be a defining moment for many. We know that it has physical and psychological symptoms and can decrease the achievement of students who are filled with potential. My hope is that you will use the strategies in this book to create a math environment that all students thrive in and transform those with math anxiety to MATH ANXIETY SURVIVORS!

I hated math most of my life, and I feared going to class every day! There are so many students who feel this way, and it is up to teachers, parents, and society to help them with this. The strategies that I have shared in this book helped me become a MATH ANXIETY SURVIVOR. This does not mean that my math anxiety is gone. It means that I now have the tools to cope when I'm having a problem. If you or any of your students suffer from math anxiety, the good news is that you are not alone. Over 90% of Americans report having negative experiences in math, but it is possible to survive this! My hope is that you will be able to implement these strategies in your math classrooms to inspire your students to be MATH ANXIETY SURVIVORS!

Chapter 4:
Additional Resources

"Education is the most powerful weapon we can use to change the world."

Nelson Mandela

The Bernard Discourse Model™

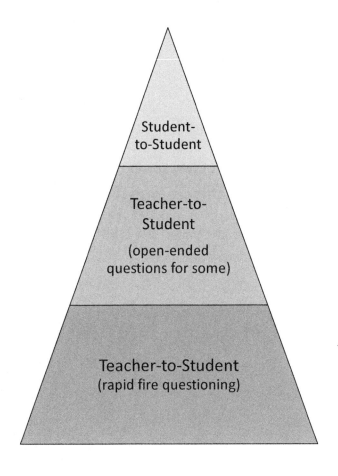

Following are some questions that promote higher order thinking and increase discourse.

Reflection and collaboration

1. What are your thoughts about what was being said?

2. How would you persuade us that you have a better method?

3. How might you argue or disagree with what we discussed?

Self-reflection

1. Why does that answer make sense to you?

2. Why didn't you attempt a different method to solve the problem?

3. Is there a way to show us exactly what you mean by that?

Reasoning

1. How might you argue against this?

2. Why did you assume this to be true?

3. Can you show us your proof for that?

Analysis

1. How would you show us the differences and similarities?

2. How many possibilities can you think of and why?

3. What patterns might direct you to a different answer?

Connections

1. How does this relate to what is going on in our daily lives?

2. What other problems fit this example?

3. What current events does this relate to?

Application

1. How can you make use of _____?

2. How could _____ be put into practice?

3. Is there another use for this?

What Are You "SMART" At?

Multiple Intelligences Survey

Directions: Circle a yes or no to each question.

1.	I like to write my own stories.	Yes	No
2.	I like to take pictures.	Yes	No
3.	I like to plant seeds and grow plants.	Yes	No
4.	I like to count.	Yes	No
5.	I like to check the weather.	Yes	No
6.	I like to talk to others about ideas.	Yes	No
7.	I like to read in my free time.	Yes	No
8.	I like to dance.	Yes	No
9.	I like to read maps.	Yes	No
10.	I like to read patterns.	Yes	No
11.	I like to take turns.	Yes	No
12.	I like to solve puzzles.	Yes	No
13.	I like to sing.	Yes	No
14.	I like to make things with my hands.	Yes	No
15.	I like to talk to others.	Yes	No
16.	I like poetry and rhyming.	Yes	No
17.	I like to be by myself sometimes.	Yes	No
18.	I like to touch different materials.	Yes	No
19.	I like to move around a lot.	Yes	No
20.	I like to play alone.	Yes	No
21.	I like to play word games.	Yes	No
22.	I like taking care of animals.	Yes	No
23.	I like to draw.	Yes	No
24.	I like to reflect on my own.	Yes	No

Math Anxiety

Areas with the most yeses reflect *which* SMART they are. For example, if they circled yes to questions 8, 13, and 16 then they are music smart.

Music	Body	Nature	Word
8, 13, 16	14, 18, 19	3, 5, 22	1, 7, 21

Math	Self	People	Picture
4, 10, 12	17, 20, 24	6, 11, 15	2, 9, 23

Music smart: This student has a high musical intelligence. They often think through rhymes and melodies. They love singing, humming, tapping their feet and hands, and playing instruments. Teachers need to offer sing-along time, opportunities to listen to music while working, and musical instruments.

Body smart: This student possesses skills with bodily and kinesthetic intelligence. The Body Smart child thinks through things that affect their body. They love running, jumping, building, gesturing, and dancing. Teachers need to give them opportunities to move, build, play physical games, engage in tactile experiences, and experience hands-on learning.

Chapter 4: Additional Resources

Nature smart: This child has naturalist intelligence. They think by observing and interacting with nature. They love the outdoors, animals, collecting, and hiking. Teachers need to give them experiences outside, field trips, and animals.

Word smart: This student has a high linguistic intelligence. They think in oral and written words. They love talking, reading, telling stories, writing, and playing word games. Teachers need to give them opportunities to read books, write stories, keep journals, and discuss stories.

Math smart: Students who are strong in this area possess skills in math and logical intelligence. These students think by reasoning and problem solving. They love math, experimenting, questioning, solving logic puzzles, and calculations. Teachers need to provide this student with things to explore and think about, science materials, manipulatives, experiments, and problems to solve.

Self-smart: These students are strong in intrapersonal intelligence. The self-smart student thinks deeply inside themselves. They love setting goals, dreaming, working quietly, and planning. Teachers need to provide this child with time alone, self-paced projects, choices, and time to reflect.

People smart: These students are strong in interpersonal intelligence. They think by sharing ideas with other students. They love leading, organizing, mediating, socializing, and communicating. Teachers need to provide group games, social gatherings, clubs, peer tutoring, and mentors.

Picture smart: This student works well with spatial intelligence. They think in images and pictures. They love designing, drawing, visualizing, and doodling. Teachers need to give this child chances to draw, books to illustrate, puzzles, mazes, and games, and use their imagination.

The Math Anxiety Bill of Rights

Source: Sandra Davis, in Doonday and Auslander (1980) Resource Manual for Counselors/Math Instructors: Math Anxiety, Math Avoidance, Reentry Mathematics

- I have the right to learn at my own pace and not feel put down if I am slower than someone else.

- I have the right to ask whatever questions I have.

- I have the right to need extra help.

- I have the right to ask a teacher or tutor for help.

- I have the right to say I don't understand.

- I have the right to not understand.

- I have the right to feel good about myself, regardless of my abilities in math.

- I have the right to not base my self-worth on my math skills.

- I have the right to view myself as capable of learning math.

- I have the right to evaluate my math instructors and how they teach.

- I have the right to relax.

- I have the right to dislike math.

- I have the right to define success in my own terms.

* * *

Helpful books for math anxiety, confidence, and study skills are available at www.brettbernard.com

Total Math Engagement by Brett Bernard (information and teaching strategies for teachers and parents)

How to Talk Math by Brett Bernard (a whole-brain workbook that teaches over 100 standards-based math vocabulary words)

Chapter 4: Additional Resources

* * *

Some helpful websites for math anxiety and math study skills

1. www.brettbernard.com
2. www.math-anxiety.com
3. www.mathacademy.com
4. www.mathplayground.com
5. www.awm-math.org
6. www.mathpower.com
7. www.math.com/students/advice/anxiety/html
8. www.worksheetworks.com
9. www.kahnacemy.org
10. www.hoodamath.com

Test taking strategies

Students should visit TestTakingTips.com.

Repetition is important in math. Students learn how to solve problems by doing them, and they should practice problems but not do it blindly.

Math Anxiety

Make sure your child or student learns how to recognize when/why they should use a specific method to solve a problem. They should work on practice problems for each topic ranging in levels of difficulty.

When practicing, they should try to solve the problem on their own first, and then look at the answer or seek help if they are having trouble.

They should mix up the order of the questions from various topics when they are reviewing, so they will learn when to use a specific method/formula.

Students should make up a sheet with all the formulas they need to know, and they should memorize all the formulas on the sheet. When they take an exam, they should write down all the key formulas in the margins. If they forget a formula during the test, they can look back at the formula.

Students should read the directions carefully and answer all parts of the question.

They should make estimates for the answers. For example, if they are asked to answer 48 x 12 =?, they could expect a number around 500.

But, if they end up with an answer around 5000, they'll know they did something wrong.

Students should show all their work (especially when partial credit is awarded) and write as legibly as possible. Even if they know the final answer is wrong, they shouldn't erase the entire work, because they may get partial credit for using the correct procedure.

When they are done, they should check every answer. If they have time, they should redo the problems on a separate piece of paper to see if they came up with the same answer the second time around.

Students should look for careless mistakes such as making sure the decimal is in the right place, that they read the directions correctly, that they copied the numbers correctly, that they put a negative sign if it was needed, that their arithmetic was correct, and so on.

Resources

Ashcroft, Mark H. and Alex M. Moore (2009). *Math Anxiety and the Affective Drop of Performance.* Journal of Psychoeducational Assessment April 13, 2009.

Beilock, S. L., E. A. Gunderson, G. Ramirez, & S. C. Levine (2010). *Female teachers' math anxiety affects girls' math achievement.* Proceedings of the National Academy of Sciences in the United States of America, 107(5), 1860-1863.

Boaler, J. (2009). *What's Math Got to Do with It?: How Parents and Teachers Can Help Children Learn to Love Their Least Favorite Subject.* New York, NY: Penguin Group.

Blatner, D. (1999). *The Joy of Pi.* USA: Walker Publishing Company.

Brenny, K. and K. Martin (2005). *1000 Best New Teacher Survival Secrets.* Naperville, IL: Sourcebooks, Inc.

Calkins, L., M. Ehrenworth, and C. Lehman (2012). *Pathways to the Common Core Accelerating Achievement*. Portsmouth, NH: Heinemann.

Cavanaugh, S. (2007). *Understanding 'Math Anxiety.'* Education Week, February 21, 2007. Retrieved from http://www.edweek.org.

Chapin, S. (2006). *Math Matters: Understanding the Math You Teach Grades K-8.* Sausalito, CA: Math Solutions Publications.

Clawson, C. (1999). *Mathematical Mysteries: The Beauty and Magic of Numbers.* New York, NY: Perseus Books Group.

Clawson, C. (1999). *Mathematical Sorcery: Revealing the Secrets of Numbers.* New York, NY: Perseus Books Group.

Cohen, D. (1988). *Calculus by and for Young People (ages 7, yes, 7 and up).* Champaign, IL: Dan Cohen The Mathman.

Delvin, K. (1998). *Life by the Numbers.* USA: John Wiley and Sons, Inc.

Furner, J. M., & B. T. Berman, (2003). *Math Anxiety: Overcoming a Major Obstacle to the Improvement of Student Math Performance*. Childhood Education, 79(3), 170-175.

Geist, E. (2010). *The Anti-Anxiety Curriculum: Combating Math Anxiety in the Classroom*. Journal of Instructional Psychology, 37(1). Retrieved from http://www.faqs.org/periodicals/201003/2011820081.html.

Horstmeier, D. (2004). *Teaching Math to People with Down Syndrome and other Hands-On Learners.* Bethesda, MD: Woodbine House, Inc.

Jackson, R. (2009). *Never Work Harder Than Your Students and Other Principles of Great Teaching.* Alexandria, VA: Association for Supervision and Curriculum Development.

Levine, M. (2013). *Teach Your Children Well: Why Values and Coping Skills Matter More Than Grades, Trophies and "Fat Envelopes."* New York, NY: Harper Perennial.

Marzano, Robert J. and Debra J. Pickering (2010). *The Highly Engaged Classroom.*

November, A. (2012). *Who Owns the Learning? Preparing Students for Success in the Digital Age*. Bloomington, IN: Solution Tree Press.

Quinn, P. (2012). *Maximum Tier 1: Improving Full Classroom Instruction.* USA: Julien John Publishing.

Ripley, A. (2014). *The Smartest Kids in the World: And How They Got That Way.* New York, NY: Simon and Schuster Paperbacks.

Rising—I attended her workshop at the National Council of Teachers of Mathematics Regional Conference in October 2013.

Siegel, D. and T. Bryson (2011). *The Whole-Brain Child: 12 Revolutionary Strategies to Nurture Your Child's Developing Mind, Survive Everyday Parenting Struggles, and Help Your Family Thrive.* New York, NY: Delacort Press.

Sparks, S. D. (2011). *Researchers Probe Causes of Math Anxiety.* Education Week, 30(31). Retrieved from http://www.edweek.org.

Tammet, D. (2014). *Thinking in Numbers: On Life, Love, Meaning and Math.* New York, NY: Hatchet Book Group.

Tobias, S. (1993). *Overcoming Math Anxiety* (Revised and Expanded). USA Haddon and Craftsman Inc.

Tough, P. (2012). *How Children Succeed: Grit, Curiosity, and the Hidden Power of Character.* New York, NY: Houghton Mifflin Harcourt.

Trapolsi, A. (2012). *Teaching Struggling Readers to Tackle Math Word Problems.* New York, NY: Scholastic.

Weimer, M. (2013). *Humor in the Classroom: 40 Years of Research.* The Teaching Professor newsletter.

Wong, H. (2001). *The First Days of School: How to be an Effective Teacher.* Mountain View, CA: Harry K. Wong Publications.

About the Author

Brett Bernard began his teaching career in 1995. He has devoted his entire career to developing and implementing strategies to engage students and bring out their strengths. Brett is a **Math Anxiety Survivor**, and he has a gift for using his struggles, failures, and experiences in math to bring out the best in all students. As a leader on the U.S.A. Unicyle Team and the 1994 World Champion of Unicycling, Bernard knows what it takes to set goals, overcome adversity, and achieve.

Brett and his wife, Allison, met as teachers when they had classrooms right next door to each other. They have two kids, Brody and Nora.

Mr. Bernard has his M. Ed in Curriculum and Instruction, is a teacher-observer, and he has coached track and cross-country running. He is a regular speaker at state and national educational conferences as well as schools across the country.

For more information and free resources to help your school achieve, please visit
www.brettbernard.com

You may contact the author at
brett@brettbernard.com

Printed in Great Britain
by Amazon